AF875419

Counting God's Blessings

A Rhyming Counting Book for Little Ones

For Kaiden and Cameron,
who are both God's blessings to me

Copyright by Carol Peterson, 2026

All rights reserved. No portion of this book may be reproduced, stored in a retrieval system, or transmitted in any form or by any means—electronic, mechanical, photocopy, recording, scanning, or other—except for brief quotations in critical reviews or articles, without the prior written permission of the publisher.

ISBN: 9781951587253

Mustard Seed Books,
an imprint of Honor Bound Books

Counting God's Blessings

A Rhyming Counting Book for Little Ones

Written and Illustrated by

Carol Peterson, MTS

When counting

blessing number

1

♥

I thank God for

yellow sun

When counting blessing number

2

I thank God for
me and you

When counting

blessing number

3

I thank God
for each
green tree

When counting blessing number

4

Jesus loves me
more and more

When counting

blessing number

5

♥ ♥ ♥ ♥ ♥

I thank God for things alive

When counting

blessing number

6

❤ ❤ ❤ ❤ ❤ ❤

I thank God
for nests
of sticks

When counting

blessing number

7

I’m happy God

is

up in heaven

When counting

blessing number

8

Thank you
for what
You create

When counting

blessing number

9

I'm thankful
for God's
BIG
design

Before I get to

blessing

10

❤❤❤❤❤❤❤❤❤❤

I want to

count them

all again

1
2
3
4
5

6

7

8

9

Counting blessing number 10

Thank you for it all...

Amen!

Grown ups understand the idea of a blessing from God. But for little ones, the word *blessing* is not one of the first words they learn or one they use often.

We might help our children understand "God's blessings" as simply gifts from Him. He gives us many kinds of gifts, such as the sun and families and everything in the world for our needs and for us to use and enjoy. Those gifts are free to us. We don't have to do anything to earn those gifts.

Here's a little rhyme you might read to your child to help explain the idea of a blessing and why God the Father, the Son (Jesus), and the Holy Spirit give those gifts to us freely.

A blessing
is a gift from He
who gives us
every gift for free.

He's God the Father,
who above
with Spirit, Son
are truest love!

Enjoy this book with your own precious blessing again and again. Your child is not only a blessing to you, but is a blessing to the world.

Made in the USA
Monee, IL
05 November 2026

67380369R00019